Smelly
Old History
Vile Vikings

D1332899

Mary Dobson

OXFORD UNIVERSITY PRESS

Oxford University Press, Great Clarendon Street, Oxford OX2 6DP

Oxford   New York
Athens   Auckland   Bangkok   Bogotá   Bombay
Buenos Aires   Calcutta   Cape Town   Dar es Salaam
Delhi   Florence   Hong Kong   Istanbul   Karachi
Kuala Lumpur   Madras   Madrid   Melbourne
Mexico City   Nairobi   Paris   Singapore
Taipei   Tokyo   Toronto   Warsaw

and associated companies in
Berlin   Ibadan

Oxford is a trade mark of Oxford University Press

© Mary Dobson 1998
First published 1998

Artwork: Vince Reid.
Photographs: York Archaeological Trust: 5, 13, 24;
Statens Historical Museum, Stockholm: 28;
C.M. Dixon: 29.

A CIP catalogue record for this book is
available from the British Library

ISBN 0-19-910494-8

1 3 5 7 9 10 8 6 4 2

Printed in Great Britain

CONTENTS

Scratch the scented panels lightly with a
fingernail to release their smell.

A SENSE OF THE PAST

Waft yourself back one thousand years to the vile Viking Age, and imagine a life of action and adventure on the salty seas. Recall the reeking raids of those vicious invaders. Savour the maggoty flesh of their mythical monsters. And if that doesn't stimulate your imagination, try poking your nose down a Viking cesspit for a whiff of their more peaceful pongs.

Of all the senses of the past, we often forget the sense of smell! This book takes you as close as possible to smelly old history. It's filled with the stinks of Vikings times, as well as a few refreshing reminders of the not-so vile Vikings.

A foul Viking warrior.

The Vikings came from Scandinavia (Norway, Sweden and Denmark) and were great sailors. They explored four continents of the globe - Europe, Africa, Asia, even America! - trading, invading, raiding and settling. At times, their actions were ferocious and they soon gained a vile reputation. But there were also peaceful Vikings who just wanted to set up home. Some of those homes have recently been found, and the things they left behind have caused quite a stink!

A vile Viking toilet.

VILE VIKINGS

Imagine a life on the open seas,
A rolling ship and a fishy breeze.
Imagine the stench of the sickly air
Filling your nose and spoiling your hair.

Now waft yourself back to those seafaring days
When Vikings stormed and ruled the waves.
Your hair is red, your sword is long,
Your breath is foul, but your will is strong!

You've set your sights on silver and gold,
To steal and loot with intentions bold.
Your journey is vile, and the sea's jolly rough,
Your shipmates are sick - is this smelly enough?

But soon your stomach heaves with delight
As you smell some sheep - hooray, land is in sight!
You order your men to make ready to go -
With one deadly swoop they deliver the blow.

You're up to your ears in vomit and mud,
Your enemies, sadly, wallow in blood.
You pull out their lungs, and you raid their store,
Then it's back to your wife with riches galore!

Now remember the Vikings were never all vile.
There's plenty of fun here to make you smile.
We'll give you a whiff of their fish and their feet,
And remind you that Vikings were sometimes quite sweet.

A strong smell of fish
hung around many
Viking homes.

A smelly old Viking sock!

5

FISH AND FUMES

Take a whiff of life in the Vikings' homeland of Scandinavia. This 9th-century settlement in Norway is full of busy Vikings at work and play.

Kleggi Horse-fly, a slave or thrall, does all the dirty work, like picking through bones. He's not too enthralled by his life! But these bones make great tools – goose legs for whistles, and tiny bird bones for needles.

Drumbr Bloodaxe ha chopped up a whale – w is being smoked for t long winter months. blubber smells vile, b provides pure oil for la *Scratch and sniff for a welcome whiff!*

Little Blubber-Feet has just slipped on to a slithering mass of eels. Unlike her brother, who is fishing in the fjord, she doesn't find it funny!

Thora is busy making cloth for clothes, sails and blankets. She has found an excellent way of getting rid of the dirt and oil – soaking the cloth in urine!

Rollo Dung-Beard is putting the finishing touches to his new longhouse – a spot of animal manure on the walls to keep them waterproof, a pig's bladder for the window, and a fresh sacrifice on the outside to please the gods.

MYTHS AND MAGGOTS

The Vikings worshipped many gods and goddesses. They also believed in lots of mythical creatures. They even had their own version of smelly old history. Viking history started at the beginning (where else?), when there was nothing but emptiness. Then an evil frost giant called Ymir emerged from the vile vapours. The first creatures slithered out of his sweaty armpit.

Ymir was slain by the god Odin and his brothers, and the world was made out of his flesh.

His blood was used for the oceans (perhaps he had blue blood).

Odin had one eye (he swapped the other for a cup of wisdom). He was the king of the gods. He was gobbled up by a wolf in one of the horrible stories of the end of the world.

His unbroken bones poked up as mountains.

His skull was the dome of the sky.

His brains floated about as clouds (not so nice when it rained).

His broken bones, teeth and bits of jaw stuck out as cliffs, rocks and stones.

While chopping up Ymir, the gods noticed that he was crawling with maggots. These turned into dwarves, elves and gnomes.

Thor was the sky god, he was very strong and controlled the weather. He rode his chariot across the sky, pulled by Toothgnasher and Toothgrinder, smashing giant snakes with his hammer to make thunder and lightning.

9

TERRIFIC TRADERS

If you were sitting in a smelly hut on a freezing winter night, listening to the eerie howls of wolves outside and the thunder of Thor, you might well start to long for adventure in other lands. The Vikings did.

From the 8th century, they began to sail the seas in search of trade and treasure. Bear skins from the frozen north were exchanged for hot spices from the east. Silk and silver, glass and wine were brought home, traded for finger-licking Viking delights made from animal bones, antlers and horns, cleaned and turned into weapons, tools, combs, and jewellery.

This Viking will trade his old skin and bones for some fancy goods from overseas.

Birka

Dublin

Jorvik

Hedeby

This Viking is sampling a glass of wine, and offering a cow horn in exchange.

The cruellest trade was in slaves.

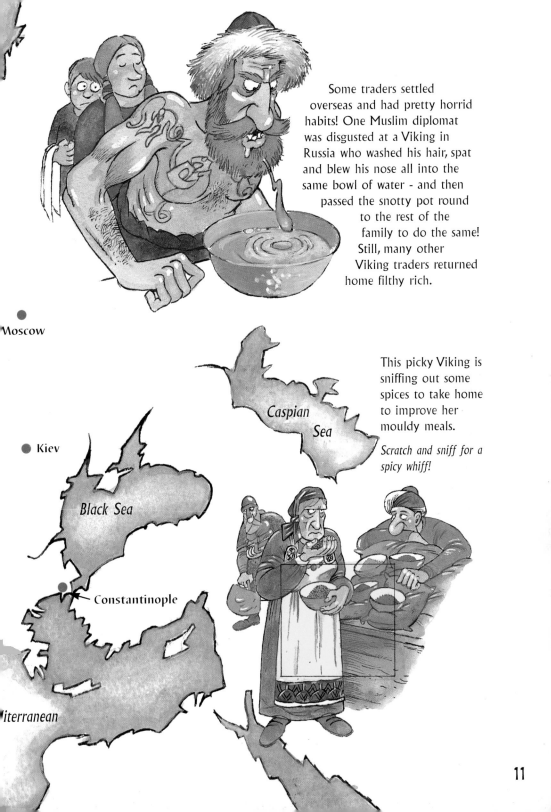

Some traders settled overseas and had pretty horrid habits! One Muslim diplomat was disgusted at a Viking in Russia who washed his hair, spat and blew his nose all into the same bowl of water - and then passed the snotty pot round to the rest of the family to do the same! Still, many other Viking traders returned home filthy rich.

Moscow

Caspian Sea

Kiev

Black Sea

Constantinople

iterranean

This picky Viking is sniffing out some spices to take home to improve her mouldy meals.

Scratch and sniff for a spicy whiff!

REVOLTING RAIDERS

Other voyaging Vikings were not so peaceful. With their fearful dragon-headed ships and deadly double-edged swords, they were the revolting raiders of many countries. Treasures were ripped from churches, lungs were ripped from enemies, and bodies of priests and peasants were trampled like dung in the streets. The smell of blood poisoned the air, sending shock waves across the world.

The very worst were the berserkers, warriors devoted to Odin. They fought like wild beasts, rolled their eyes, and howled like angry animals. They went into battle unprotected, wearing only bear skins, or even bare skins. It is thought they chewed poisonous toadstools which made them go berserk!

Ivar the Boneless was a great Viking hero. He invaded Anglo-Saxon Britain in 865 with a great army, including berserkers. The Vikings took over large parts of Britain after this, but the Saxon King of Wessex, Alfred the Great, also did a great job to prevent a complete Viking walkover.

Double-edged swords, with blood channels running down each side, were given gripping names like 'leg-biter'.

13

ᚠ

FOUL FEASTING

Fighting and feasting went together in the Viking age. This lavish feast has been laid on to celebrate the return of the victorious warriors. Everyone is invited — including a few dead relatives, who have their own special table and are welcomed back from the 'outside' with a fresh steam bath to remove their grave remains.

SLIMY SMOKED SEAGULL ON THE BONE

SLABS OF SIZZLING SEAL STEAK ON STICKS

RAW SAUSAGES MADE FROM LARD, BLOOD & REINDEER MEAT, SMOKED & SPICED WITH CELERY SEED & LEEKS

PICKLED PUFFIN PIE AND SALTY SEAWEED STEW

Sour milk and strong mead (a powerful brew made from honey) are drunk from cattle and reindeer horns – everyone's very thirsty because the food is so salty. Lots of mouldy gunge gets trapped in the bottom of the horns, but by the end of the feast these revolting revellers are just too merry to care about that.

Scratch and sniff the delicious dish.

ODOROUS ORDEALS

The Vikings had some vile ways of dealing with their enemies. Many odorous ordeals were described in sagas (stories) or poems. These were not written down until much later in Viking history, so it's not always easy to sniff out the truth.

Thingy the Evil is accused of being a witch and is brought before the Thing (a get-together of Vikings) for a trial by ordeal. She has to pull out a handful of burning stones from a cauldron of boiling water and hold them for a few terrible seconds. If, after four days, her hand is festering, she will be outlawed so that anyone can murder her without penalty.

Bersi the Bold and Olaf Hoochoose are jarls (chieftains) of two Viking clans. They are arguing over who owns a pile of bones, so will resolve it by fighting a duel. When blood is spilled, the duel is over and the wounded man has to pay up in silver.

Thrand the Stinking has been wounded on the battlefield. The arrow has gone deep and can't be found. Fredda Bignose is stuffing him with a mixture of leeks and herbs, and sniffing his open wounds. The one that smells of leeks contains the deadly weapon.

A Viking chieftain, Thorgrim Scarfoot, and his clan are enjoying sacrificing a few foes to their gods. Behind the bushes a captured enemy is being spread-eagled by Ragnar Claw-Face. In this most vile practice, a cut is made down the spine of the living victim, the ribcage is opened sideways and the lungs are ripped out like an eagle's wings.

FIERY FUNERALS

Viking life didn't end with death. If a warrior was lucky and died on the battlefield, he would hope to go to Valhalla, Odin's splendid hall for the dead. Here, according to Viking beliefs, he would enjoy endless feasting and fighting.

Thorfinn Skull-Spliter, a famous chieftain, has died heroically on the battlefield. A charming corpse chooser, or Valkyrie, selects him for the after-life.

Skull-Spliter will sail with his horse, dog, weapons, and one of his finest slaves to another world.

Skull-Spliter joins his friends for a deliciously aromatic after-life in Valhalla. Sarimmer the succulent pig is slaughtered daily, roasted and eaten, and rises up grunting again next morning. Fresh mead from the udder of Odin's goat is drunk from the skulls of enemies.

Viking warriors who had the misfortune to die peacefully in their beds would expect a much less fragrant after-life in the underworld called Nifelheim.

Ketill Flatnose has died peacefully in his sleep. He is doomed forever.

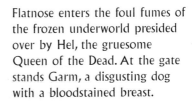

Flatnose enters the foul fumes of the frozen underworld presided over by Hel, the gruesome Queen of the Dead. At the gate stands Garm, a disgusting dog with a bloodstained breast.

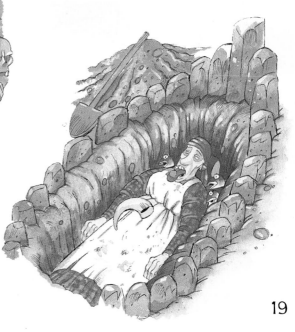

Many Vikings probably ended up in a hole in the ground with just a few bits and pieces for their after–life. Helgi has a coin in her mouth, to pay the ferryman to take her to the other side.

A FRESH START

The Viking age wasn't all doom and gloom, and the Vikings weren't vile all the time. The bravest and most daring set off on amazing voyages of exploration. Follow the adventures of Erik the Red and his crew to enjoy a fresh (well, fairly fresh) view of the smelly old Viking world.

Viking explorers have settled in Iceland. Erik the Red, a man of fiery temper, kills some of his neighbours and is made an outlaw.

To save his skin, he must pack his bags and flee.

Erik and his crew brave the bitter Arctic winds and the terrible stench of seasick sailors. Eventually, the welcome smell of reindeer dung wafts their way - they are near land.

Erik claims the land and calls his frozen find Greenland, hoping that others will come if it has a nice fresh name. And they do. In 985 a thousand settlers and their animals squeeze on to 25 ships. Only 14 ships survive the smelly trip. But others soon follow, and Erik becomes the great Viking chief of Greenland.

Not to be outdone by his old dad, Erik's son, Leif Eriksson, goes one better. He sails the ocean blue in 1002 and beats Christopher Columbus to America by 490 years! He finds a land of plenty and calls it Vinland ('Wineland'). Leif's settlement in Old Vinland was recently found in Newfoundland!

21

PEACEFUL PONGS

Just what was it like to live in a Viking town? Thanks to the amazing discovery of masses of remains under the streets of the city of York, in northern England, we now have a pretty good idea — it was extremely smelly! Over 10,000 Vikings lived in Jorvik — take a look, and even a whiff, of some of them here.

Frigga Knockknees is jostling with busy shoppers eager to snatch a fresh catch of eels.

The River Ouse is oozing with unmentionable filth, but busy with trading boats.

It's so dark in Astrid One-Eye's house that she doesn't notice the bugs on the muddy floor. She's spinning wool to make socks.

Bjorn is enjoying a little privy privacy.

Scratch and sniff for a peaceful whiff!

VIKING VERMIN

Digging up old Viking sites has caused quite a stink! Historians have had a rare chance to take a close look at the insides of the Viking world. They have discovered that:

Viking toilets were the absolute pits - just stinking holes in the ground, covered with planks like this one. Vikings used moss, bits of old rag or oyster shells as loo paper.

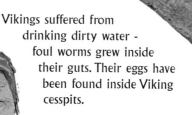

Vikings suffered from drinking dirty water - foul worms grew inside their guts. Their eggs have been found inside Viking cesspits.

Viking vermin spread far and wide - plagues of rats and mice landed on the shores with the invaders.

Vile plagues and putrid pestilences often wiped out more people than the battles. Many died of a disease called the skit, or bloody flux. They thought it was caused by demons breathing venom and fire. Now we know that it was caused by the foul conditions which allowed the rapid spread of germs.

There were no trained doctors in Viking times, and revolting cures included bull's gall, goat's urine mixed with honey, and dog's brain. In spite of all this, the Vikings were very proud of their appearance:

They worked hard to remove the weekly grime and sweat in the Saturday steam bath. The Viking day for Saturday means 'bath-day'.

Evil-smelling soap was made from wood ashes, nettles or cow dung.

Fashion-conscious Vikings had regular grooming sessions to remove lice and fleas.

In Britain, the smelly Saxons complained that the vile Vikings attracted all the pretty women because they combed their hair, had baths and changed their clothes so often!

25

SMELLS AND BELLS

The Vikings' gods eventually came under attack from crusading Christians. Vikings living in England and France heard about the new religion first, but in the end even those back home were forced to convert — by fair means or foul. Once King Olaf Tryggvason of Norway had taken the plunge, he was ruthless in forcing his people to do the same. If they refused, their eyes were burnt out and their tongues were cut off — not surprisingly, most of them agreed!

These freshly converted Vikings face another problem. The bells are ringing out to welcome them to church, but inside there are an awful lot of unwashed bodies. Fortunately, the Christian church has a solution for body odours - sweet-smelling incense is wafted around to purify the atmosphere.

Scratch and sniff here, for a rare Viking perfumed pong.

By the 11th century, the bloody Viking age of invading and raiding was grumbling to an end. But Viking descendants were all over the place by this time. Even William of Normandy, who conquered England in 1066, had Viking blood in him. You may even have some yourself!

Smelly old history doesn't end in 1066. Wait till you smell that *Medieval Muck*!

GRAVE REMAINS

Viking history was not written down until the 12th and 13th centuries, and by then it was probably pretty inaccurate, so reading about the Viking past can be difficult. Some of the best pieces of Viking evidence are the actual grave remains that accompanied the dead to the after-life, and the bits of human muck that accompanied the living to the cesspit.

Earspoons for cleaning out ears.

In Oseberg in Norway this magnificent ship was found buried in silt and slime. It contained the bones of two women: a queen and her sacrificed slave. Everything that was needed for the after-life was buried with the lady, including peacocks, dogs, a pair of special shoes for her deformed feet, a manure fork for spreading muck, and spices to keep her apples and bread fresh! She must have expected to be forever on the move, as she also took with her a wagon and four sledges, two oxen, ten horses, two tents and a camp bed!

The huge rubbish tips at Viking settlements have also preserved some whiffy specimens. Lumps of human excrement, a fleshless fishbone or the odd sock can tell us heaps about smelly old history. So much rubbish was dumped in Jorvik during the 10th century, the ground level rose by 2.5 cm every year.

This beautiful carving decorated a bucket in the Oseburg ship-burial.

PUNGENT PUZZLES

REEKING RUNES

The Vikings had an alphabet, called the futhark, which had 16 characters called runes.

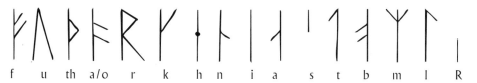

f u th a/o r k h n i a s t b m l R

Can you translate the words in these runes?

Viking children didn't go to school or learn to read or write, so they would have had a hard time solving this pungent puzzle!

NASTY NICKNAMES

Nasty nicknames were given to vile Vikings by later writers. The Vikings would probably turn in their graves if they knew how rude historians have been about them! Find five nasty nicknames in this word search.

WHACKY WEEKDAYS

The gods Tiw, Woden, Thor and Frey gave their names to four days of the week – guess which ones.

```
F  R  D  A  J  P  B  R  T  K
I  K  B  O  N  E  L  E  S  S
N  S  E  M  Q  T  O  A  X  R
G  F  L  A  T  N  O  S  E  C
L  C  P  O  R  I  D  N  E  J
D  U  N  G  B  E  A  R  D  Y
R  I  T  L  I  C  X  G  K  P
X  S  A  Z  J  O  E  V  I  L
```

GLOSSARY

berserker	A ferocious Viking warrior.
cesspit	A deep pit where rubbish and sewage were thrown.
diplomat	Someone who travels to a foreign country to represent their own country.
incense	A substance that makes a spicy smell when it is burned.
jarl	A Viking chieftain.
mead	An alcoholic drink made from honey.
Nifelheim	In Viking mythology, the underworld.
pestilence	A plague.
rune	A character in the Viking 'alphabet'.
saga	A Viking story or long poem.
Scandinavia	The countries of Sweden, Denmark, Norway and Finland.
thrall	A slave.
Thing	A Viking assembly.
Valhalla	In Viking mythology, the hall where heroes who have died in battle feast with Odin.
Valkyrie	In Viking mythology, a handmaiden of Odin who carries dead warriors to Valhalla.
vermin	Animals, birds or insects that carry diseases or cause damage to food.

INDEX

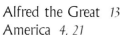